Around and About
The sky above us

Kate Petty
and Jakki Wood

Barron's

First edition for the United States, Canada, and the Philippines published 1993 by Barron's Educational Series, Inc.

© Copyright by Aladdin Books Ltd 1993

Designed and produced by
Aladdin Books Ltd
28 Percy Street
London W1P 9FF

All inquiries should be addressed to:
Barron's Educational Series, Inc.
250 Wireless Boulevard
Hauppauge, NY 11788

International Standard Book No. 0-8120-1234-8

Library of Congress
Catalog Card No. 92-30582

Library of Congress Cataloging-in-Publication Data

Petty, Kate.
  The sky above us / Kate Petty.
    p. cm. – (Around and about)
    Includes index.
    Summary: From their balloon, Harry and Ralph find out about the various layers of the Earth's atmosphere, snow, rain and other types of weather, and such meteorological devices as weather balloons and weather satellites.
ISBN 0-8120-1234-8
1. Atmosphere, Upper–Juvenile literature. 2. Troposphere–Juvenile literature. 3. Weather–Juvenile literature. 4. Meteorology–Juvenile literature. [1. Atmosphere, Upper. 2. Weather. 3. Meteorology.] I. Title. II. Series: Petty, Kate. Around and about.
QC879.15.P48 1993
551.5–dc20   92-30582   CIP   AC

Printed in Belgium

3456 4208 987654321

Design David West Children's Book Design
Illustrator Jakki Wood
Text Kate Petty
Consultants Keith Lye B.A., F.R.S.G., Eva Bass Cert. Ed., teacher of geography to 5-8 year-olds

Contents

The air above us

Harry and Ralph load the basket of their balloon with plenty of fuel. They want to climb high in the sky today to explore the air above them.

There is a layer of air all around the planet. It is called the atmosphere. Without the atmosphere, we would have no air to breathe and no protection from the Sun.

Going up

A big burst of flame from the heater and the hot air inside the balloon lifts them off the ground.

Up they go, above the houses…

and the trees…

above the town…

above the hills…and mountains.

The air is thinner the higher they go. Harry and Ralph have oxygen packs just in case they have trouble breathing.

Brrrr... it's chilly up here!

How do I look?

It's colder higher up too. Harry has some extra blankets.

Harry and Ralph can't go any higher in their balloon but they are still in the lowest part of the atmosphere, called the troposphere. The troposphere goes up for about 5 to 11 miles (8 to 18 kilometers).

Above the weather

Harry and Ralph are about 2½ miles (4 kilometers) off the ground. Their balloon is buffeted about by the wind. Dark storm clouds are approaching. Harry wishes they could go up, to the stratosphere, where they would be above the weather.

There's another balloon up there.

It's a weather balloon with no people on board, so it's all right to go very high.

Troposphe

Earth

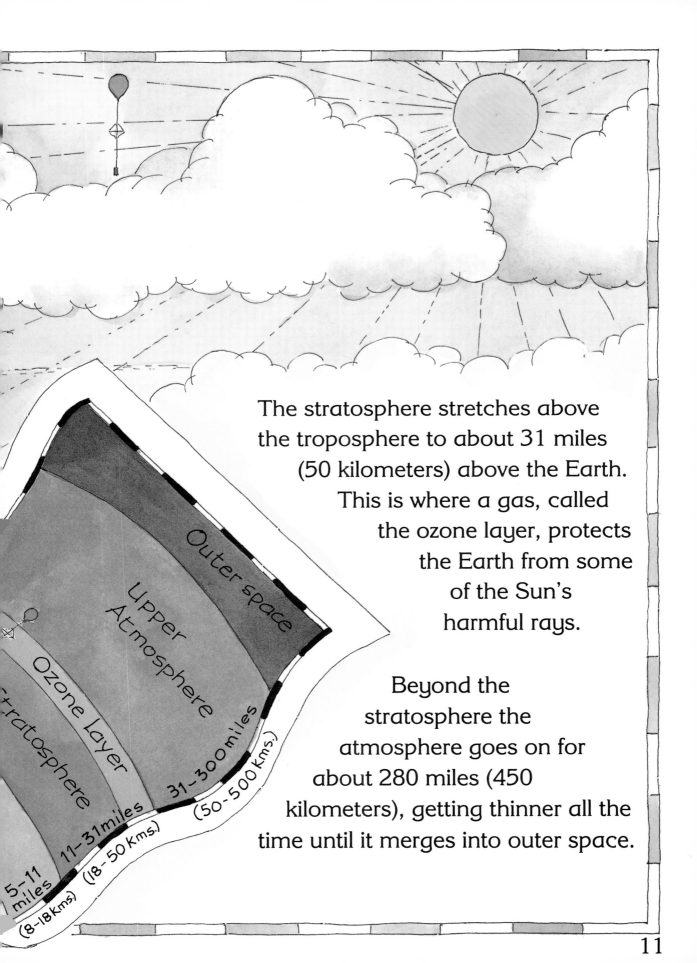

The stratosphere stretches above the troposphere to about 31 miles (50 kilometers) above the Earth. This is where a gas, called the ozone layer, protects the Earth from some of the Sun's harmful rays.

Beyond the stratosphere the atmosphere goes on for about 280 miles (450 kilometers), getting thinner all the time until it merges into outer space.

Outer space

Upper Atmosphere

Ozone Layer

Stratosphere

31-300 miles (50-500 kms.)

11-31 miles (18-50 Kms.)

5-11 miles (8-18 Kms.)

Clouds

It's time for the picnic. Harry pours two cups of hot chocolate from the thermos. Little clouds of steam rise from their cups.
Ralph is looking for the cookies, when suddenly –

they are in the middle of a cloud.

I can't see a thing.

Those water droplets are ganging up on us!

The cloud is like the steam. Fog and mist are clouds on the ground.

The air contains a lot of water vapor. This is the invisible gas that comes off the surface of water when it is heated. As the air rises to a cooler level, the water vapor in it turns back into tiny droplets of water. The droplets are so small and light that they float about in the air. Millions of droplets form clouds.

Next time someone boils a kettle in your house, stand well back and watch the steam. What happens to the window nearby?

Rain, rain

Ralph is beginning to think this trip was a big mistake.

Harry puts on his slicker. This next cloud is definitely a rain cloud.

Inside the cloud the tiny droplets of water are moving about, bumping into one another and forming bigger and bigger drops. When the drops get fat and heavy, they fall to the ground.

There's no getting away from the water cycle.

Harry looks down at the river running into the sea below and thinks about all this water going around and around.

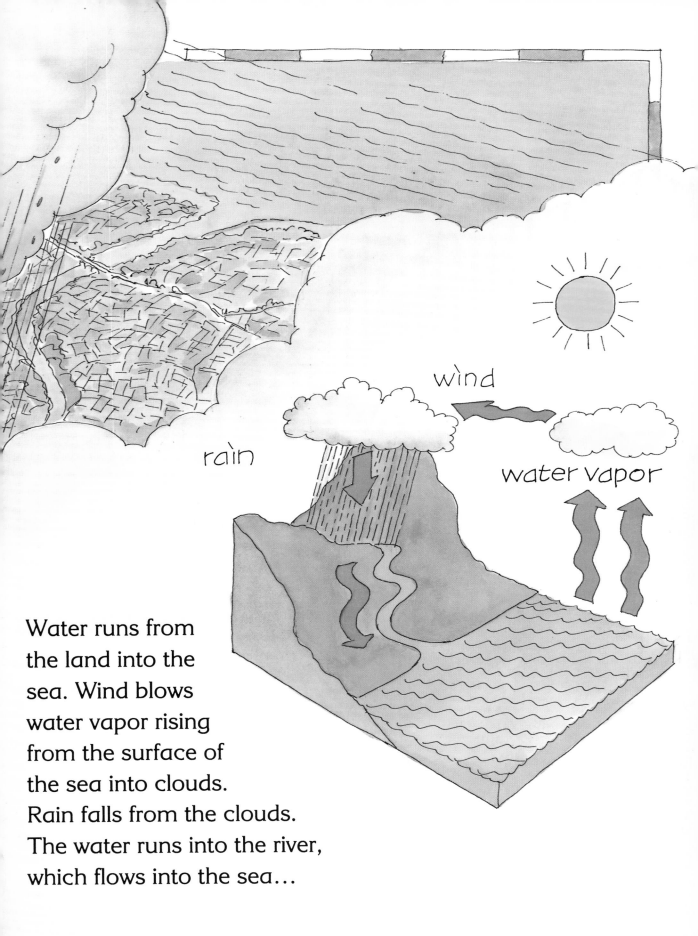

wind

rain

water vapor

Water runs from
the land into the
sea. Wind blows
water vapor rising
from the surface of
the sea into clouds.
Rain falls from the clouds.
The water runs into the river,
which flows into the sea...

The Sun

Thank goodness! The Sun has come out. In fact, the Sun never goes away. During the day it can be hidden by clouds. At night we can't see it because our globe is turned away from the Sun. In winter it seems less strong because our part of the globe is tilting away from it.

 Sun

Earth travels around the Sun.

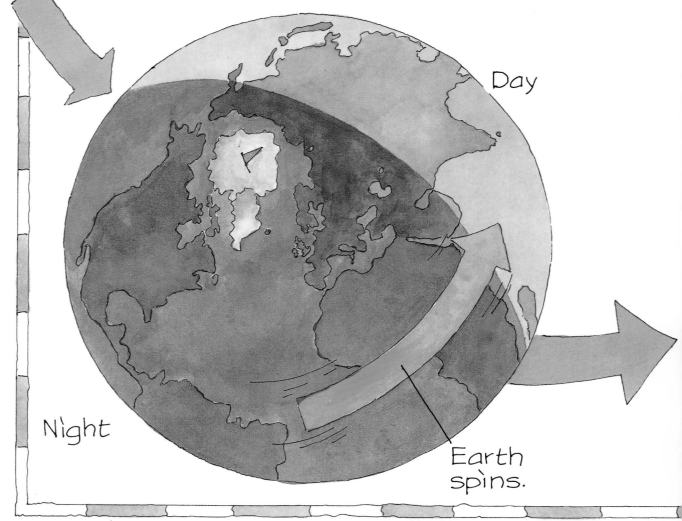

Day

Night

Earth spins.

But the Sun is always there. We get light, warmth, and life from this ball of flaming gas that is about a hundred times bigger than the world and 93 million miles (150 million kilometers) away.

Sun and rain together make rainbows.

The white light from the Sun is split into seven main colors by the raindrops.

Heat from the Sun warms the ground rather than the air. Warmth from the ground heats the air above it.

Sea breezes

Harry and Ralph bring the balloon down near the sea. The balloon sinks as the air inside it cools.

It's windy by the sea. Harry knows that it's air moving about. He asks Fred about sea breezes.

"Air is like water," says Fred. "It flows from places where there is a lot of air (called high pressure) to places where there is less air (low pressure). Low pressure can happen when hot air rises – like it does over the land on a hot day. Then air from a high pressure area – like the air over the cool sea – blows in to fill its place."

All over the world, winds are made by air from high pressure areas blowing to low pressure areas.

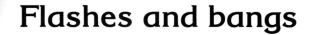

Flashes and bangs

Remember those rain clouds? The wind has blown them toward Harry and Ralph. They find shelter from the storm under Fred's boat.

FLASH! There goes the lightning! The lightning is a huge electric spark produced by all the activity inside a big storm cloud.

The spark heats up the air, which expands with a big BANG!, and that's the thunder that Harry and Ralph hear shortly afterwards.

More storms

Ralph covers his ears. He doesn't like storms. Fred says he's lucky not to be at sea!

A really violent storm that builds up over tropical seas is called a hurricane. With winds of over 125 miles (200 kilometers) an hour, hurricanes can cause terrible damage.

Another type of storm is called a tornado. Tornadoes are small, violent whirlwinds that suck up and destroy almost everything in their paths.

Many tornadoes happen in the midwestern United States.

Like Kansas, where Dorothy lives in *The Wizard of Oz?*

Snow

Ralph's favorite kind of weather is SNOW! But it's the wrong time of year for snow. It has to be really cold for the water droplets in a cloud to become ice crystals. The crystals join together to form snowflakes.

There can be snow all year-round on high mountain tops and at the North and South Poles. In other places, snow usually melts when spring comes around.

What's your favorite kind of weather? What's your favorite time of year? What is the weather like in spring, summer, winter, and autumn where you live?

The weather forecast

Harry and Ralph watch the TV to find out what kind of weather they are going to have tomorrow.

Weather forecasters get their information from weather stations all over the world. Some are on land, others are at sea. Weather balloons and aircraft carry instruments in the sky. Weather satellites orbit the Earth.

Details of temperatures, wind speed and direction, sunshine, how much water there is in the air, rainfall, and air pressure are fed into a computer. The computer displays the results as a map, like this.

Using these maps, a forecaster can work out weather patterns for the next few days.

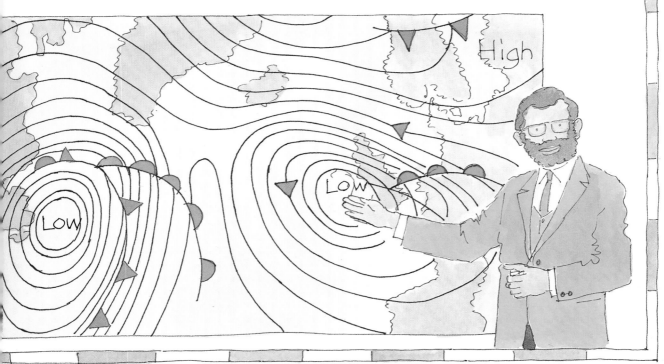

A weather station

Harry and Ralph have their own weather station at school. At the same time each morning and afternoon, they record the temperature, the rainfall, and the direction of the wind. They make a chart from all the information.

The wind vane is made from a pencil and some plastic-covered card.

There is a thermometer for measuring the temperature in the shade.

The rain gauge is made from two halves of a plastic bottle and a ruler.

It's nice out today.

And the wind is blowing from the south-west.

Index

This index will help you to find some of the important words in the book.